Let's go to school!

Could you go on a bus?

Yes, you could.

She could bike to school.

She'll be at school soon!

He goes to school at home.

What if you brought a box?

These kids took a ship to school!